D0940030

Agent H2O Rides the Water Cycle

Dedicated to fellow desert dwellers
who are fans of H2O

Rita Goldner

ISBN: 978-0-9978298-0-8
Written and Illustrated by Rita Goldner

The text is set in Papyrus
The illustrations are rendered using computer software
10 9 9 7 6 5 4 3 2 1 First Edition
Children, Conservation, Ecology, S.T.E.M.

Hi, I'm Secret Agent H2O. I've been hanging around for millions of years, since before dinosaurs roamed the earth. My life is crammed with adventure, travel, important missions, and a few a disguises.

Join me on a vital mission: to *hydrate* people, animals and plants. Today I have a great disguise for a secret agent. I'm invisible!

The sizzling heat turns me into *vapor*. You can't see it. I call this sneaky prank *evaporation*. Let's jump on the water cycle and explore this wild ride!

A bunch of other drops *evaporated* too, so there's lots of *vapor* up here. It's chilly. Time for *condensation* from an invisible *vapor* disguise back to a tiny drop.

I join other drops to make some puffy clouds. Wow, what a view!

When I find tiny pieces of dust and bacteria in the clouds, I hang on and float around with other little drops. Makes it easier to team up and make bigger drops.

Wheee! We rain down, and call this roller coaster ride *precipitation*.

Hey, let's do it again, but on a different route this time. I EVAPORATE into my vapor disguise, and then detour to a colder place.

The drop in temperature changes my shape from *vapor* into liquid. When the air hits freezing temps, I'm a solid ice crystal. Brrrr!

I almost always do the shape change in these three steps. In warm weather I melt from ice crystal to a drop, then evaporate into *vapor*.

In cool weather, I usually do the three steps backwards. I change from vapor to drop, and then I freeze into an ice crystal.

Once in a while I take a crazy shortcut and skip the middle step. This can only happen in two events: when a freezing cold day turns suddenly hot with a dry wind,

OR when a warm day gets blasted by a super cold wind.

During the fast warm-ups I go from ice straight into vapor. This stunt is SUBLIMATION.

During the fast cool-downs, I do the wacky short cut backwards. I slip from vapor straight into ice, and call this trick DEPOSITION.

Now where's a cool place for an ice crystal to hang out?

Wheeee!

YIPPEE! Secret Agent H2O conquers the ski slopes! Just don't get flattened by a snowboard.

Below, I spy desperate plants and animals in the valley. I'd better melt into liquid. Let's race downhill to the rescue!

This plant looks ready to dry up and blow away. Time to squeeze into a root and bring it back to life.

Tough work, but it's done. I seep back out in my vapor disguise. From here it's a cinch to condense into a dew drop and explore for

YIKES! Animals get thirsty, too. They cant live more than a few days without water.

I slip-slide down this coyote's throat to give him a cool drink.

Whew. Glad that job is over. Uh oh. I sense some trouble ahead on our mission.

We'd better investigate.

Like all secret agents, I have an evil villain chasing me. His name is Scummy Pollution. Gross! This dirty guy can't get rid of me.

After all, I've been around for millions of years. He wants to slow down our mission with this yucky mess. NO WAY! If that happens, people, plants and animals are in trouble. Wait, I've got a foolproof escape plan. Follow me!

It's an *aquifer*. Let's dive in. We can trickle down through the sand and gravel to ditch Scummy.

fine sand and clay (silt)
sand and gravel
aquifer

Now ooze along the underground river. The grains of sand and gravel scrape this grimy guy away.

We made it, I'm fresh and clean again. Wonder what we'll find in the river above. Let's seep up

Surprise! It's the desert river where we started. Thanks for coming along. You're welcome anytime.

I'll start a new rescue mission on the water cycle tomorrow. Keep an eye out for me. Agent H2O rides again!

Fun New Words

Aquifer: Underground layer of rock, sand, and gravel

Condensation: Vapor changes to liquid

Deposition: Vapor changes to solid without becoming liquid first

Evaporation: Liquid changes to vapor

H2O: Chemical name for water

Hydration: Provide water to a person, animal, or plant.

Percolation: Water filters down through gravel, sand, and rock

Precipitation: Rain or snow falls to earth after condensing from vapor

Sublimation: Snowflakes and frost change to vapor without becoming liquid first

Vapor: A gas; it's invisible

Online References:

Water Science School - USGS

The Water Cycle Educational Video for Kids

The Water Cycle for Kids - How it Works - Diagram & Facts

Sublimation and the Water Cycle USGS

The Fundamentals of the Water Cycle

Water Cycle - National Geographic Kids

Kidzone.ws/water

Parent/Teacher Guides

RitaGoldnerBooks.com has a science/animal blog for kids, a free Water Cycle card game to cut out, coloring pages, and a kids' newsletter.

Made in the USA
Middletown, DE
11 March 2021

35388795R00020